Flamboyant Blooming Trees

THE AUTHOR

Dr. Shobha Nagnur having completed her Bachelor's degree in Home Science, pursued her higher studies, and earned both masters and doctorate in Agricultural Extension from the University of Agricultural Sciences, Dharwad.

She is a nature lover and a garden enthusiast with photography as a hobby. An avid traveler, both professional and personal journeys have taken her to nooks and corners of the country. This has lent her an opportunity to expand the horizons of her love for nature as a hobby into a more professional appreciation of the diverse and rich flora of India. She has used these frequent sojourns to capture through her lens, the myriad play of the plant world.

She has been communicating to the general readers, the complex web of nature and environment by articulating in simple and readable form through her regular writings in newspapers and magazines. Of the three books that she has planned, the first one on climbers has already been published. This one on 'Trees' is shared with the nature loving readers, as a second one of the tri-series.

Presently, she is working as Dean of the College of Community Science, University of Agricultural Sciences, Dharwad.

She lives with her family in Dharwad, a city in Karnataka located at the cusp of the hills and plains. Situated agro-ecologically in the transition zone between wet and dry regions, Dharwad is home to a kaleidoscope of flora. Her sprawling garden with flower and fruit trees amidst the concrete jungle stands as a testimony to her passion for gardening.

Flamboyant Blooming Trees

Shobha Nagnur

2022

Daya Publishing House®

A Division of

Astral International Pvt. Ltd.

New Delhi – 110 002

© 2022 AUTHOR
ISBN: 978-93-90435-36-4 (HB)

Publisher's Note:

Every possible effort has been made to ensure that the information contained in this book is accurate at the time of going to press, and the publisher and author cannot accept responsibility for any errors or omissions, however caused. No responsibility for loss or damage occasioned to any person acting, or refraining from action, as a result of the material in this publication can be accepted by the editor, the publisher or the author. The Publisher is not associated with any product or vendor mentioned in the book. The contents of this work are intended to further general scientific research, understanding and discussion only. Readers should consult with a specialist where appropriate.

Every effort has been made to trace the owners of copyright material used in this book, if any. The author and the publisher will be grateful for any omission brought to their notice for acknowledgement in the future editions of the book.

All Rights reserved under International Copyright Conventions. No part of this publication may be reproduced, stored in a retrieval system, or transmitted in any form or by any means, electronic, mechanical, photocopying, recording or otherwise without the prior written consent of the publisher and the copyright owner.

Published by : **Daya Publishing House**®
A Division of
Astral International Pvt. Ltd.
– ISO 9001:2015 Certified Company –
4736/23, Ansari Road, Darya Ganj,
New Delhi-110 002
Ph. 011-43549197, 23278134
E-mail: info@astralint.com
Website: www.astralint.com

Digitally Printed at : **Replika Press Pvt. Ltd.**

FAMILY
Fabaceae

NATIVE
Australia

FLOWERING TIME
June –July and October –November

SCIENTIFIC NAME:

Acacia auriculiformis

COMMON NAMES :

Ear leaf acacia (English); Dal moth (Hindi); Bili Jali (Kannada); Akashmoni (Bengali); Kaththi, Karuvel (Tamil); Minnumaanu, Kondamanu (Telugu)

THE ear leaf Acacia not really grown for its flowers, bears spikes of pretty & elegant, yellow and white blooms. During bloom time the yellow-orange /white flower spikes are heavily interspersed with the leaves. The spikes are seen in drooping clusters at the end of the branches. These spikes which are the size of a finger, both in length and width are usually seen in pairs. Each spike contains at least 100 tiny fragrant flowers.

The evergreen tree produces smooth and slightly curved leaves which can be easily confused with that of the Eucalyptus (Nilgiri) tree.

The tree makes good firewood, producing good quality charcoal. The wood is also used for making paper, furniture and tools.

It is a fast growing tree and being hardy, it can be grown with little or no care. It's medium size, ornamental flowers & leaves, as well as shade providing quality render it suitable for planting along roadsides as an avenue tree within the city limits.

Propagation is by seeds which are produced in profusion. In nature the seeds are dispersed by birds.

FAMILY
Mimosaceae

NATIVE
Indian sub continent and West Asia

FLOWERING TIME
June –July and October- November

SCIENTIFIC NAME:

Acacia nilotica

COMMON NAMES :

Gum Arabic (English); Kikar (Hindi); Kari Jali (Kannada); Babul (Marathi); Babaria(Gujarathi); Akashmoni (Bengali); Nalla tumma (Telugu); Karivelam (Malayalam)

Blooming twice a year, the flowers of this tree are not discernable, but a close observation will show the tree being dotted with bright yellow fluffy flowers. Appearing in clusters of 2-6, about 1- 1.5 cm in diameter they look like miniature badminton balls.

The dark green leaves with numerous leaflets are delicate looking and feathery. At the base of the leaves is a sharp thorn which hardens as the leaves mature.

Apart from the different parts of the plant being used in traditional medicine, the tree has multiple other uses. While pods, seeds and young branches make good animal feed, the sharp thorns are used as eco- friendly paper pins, twigs are used as tooth brushes and the edible gum collected from the bark is used in the food industry.

The reasons why farmers wish to have a few of these trees on the farms is that the hard and durable wood is most suitable for making agricultural implements. The plant is known to fix nitrogen in the soil making the soil fertile.

It is a species that requires no care at all and can be used to regenerate waste lands as the seeds can germinate within two weeks.

FAMILY
Cornaceae

NATIVE
India

FLOWERING TIME
February-March

SCIENTIFIC NAME:

Alangium salvifolium

COMMON NAMES :

Sage leaved Alangium, Stone mango (English); Nikochalia, Deerg hakeela (Sankskrit); Ankol, Dhena (Hindi); Ankola (Kannada); Akorkanta (Bengali); Alinijil (Tamil); Urgu (Telugu); Arinji (Malayalam)

This small tree, more popular for its medicinal properties, bears fragrant greenish white flowers with six delicate, filament like petals. The petals bend backwards exposing the multiple stamens and a central stigma that stands out. During bloom time, the entire tree is bereft of leaves and one can see the white flowers interspersed with prominent, longish, green colour buds. Leaves are simple, oblong shaped

Round violet - purple coloured fruits follow the flowers. They are edible, with a sweetish taste. The fruits fall to the earth just in time to sprout with the first monsoon rains

This tree is considered a holy tree and so is seen in the vicinity of temples. People worship the tree believing that it would fulfill their wishes. The tree is highly valued for its medicinal properties. All parts namely; roots, leaves, stems and bark are traditionally used for curing diseases and in emergencies like poisoning and snake bites. Wood is used in making musical instruments, for furniture and in construction.

Apart from temple complexes this tree is seen in wastelands and along roadsides. It is suited for planting along field borders and avenues.

Propagation is by seeds which are often dispersed by birds

FAMILY
Fabaceae

NATIVE
Tropical Africa, Asia and Northern Australia

FLOWERING TIME
March to May

SCIENTIFIC NAME:
Albizia lebbeck

COMMON NAMES :

Flea tree, Fly wood tree, Woman's tongue tree, East Indian walnut (English), Goddahunsemara, Bagemara (Kannada), Siris (Hindi), Kothiya, Koroi, Sirish (Assamese), Vaga, Korimthakara, Kattuvaka(Malayalam), Siridam, Vagai (Tamil) and Khok (Manipuri)

When in bloom, clusters of light green, white powder puff like flowers dot the wide canopy. Each cluster contains a central flower and upto 20 circlets of flowers. The fallen flowers form a green carpet below the tree. The short lived flowers attract a lot of bees and butterflies.

The flowers on withering are soon followed by flat seed pods about two inches wide and six inches long. Initially green, the pods dry up and take a yellow colour. By November–December the entire tree is covered with the seed pods. Each pod holds 6-12 loose seeds. Even with the slightest of the breeze the pods shake and the seeds within make a rattling sound, giving the tree the epithet of 'Woman's tongue tree'.

The tree in addition to its medicinal properties also produces good quality wood. Since the tree is quick growing, ornamental and shade providing, it is commonly grown along roadsides throughout the country.

Propagation is by seeds.

FAMILY
Caesalpiniaceae

NATIVE
Central America

FLOWERING TIME
March- June/ End of the year

SCIENTIFIC NAME:
Albizia saman

COMMON NAMES :
Rain tree, Monkey pod tree, French tamarind (English); Gulabi siris, Vilaiti siris (Hindi); Male-mara, Chandina-mara (Kannada); Gulabi siris, Vilaiti siris (Marathi); Bilitis siris (Bengali);Amaivagai, Thoongumoonji-maram (Tamil); Dorisena (Telugu); Urakkamthuunga-maram, Mazha-maram (Malayalam)

The flowers of this tree can easily go unnoticed because they are perched high on top of the canopy. The best view of this wide-spread tree, is when one looks down on the canopy from atop rather than viewing from below. The flowers are borne in powder puff like clusters, similar to that of the *A. Lebbeck* except that they are a pink in colour.

With a height of up to 80 ft and spread of 100 ft, this tree is an ideal shade tree in most tropical countries. The compound leaves fold themselves in cloudy weather forecasting rain and hence the common name "The Rain Tree".

The fruits when mature are brown pods, about 1 inch wide and 6-8 inches in length, with small seeds within. The flesh of the fruit is sticky and so children crush the pods to shape them like balls, which when dry can be as hard as a cork ball. The fruits are relished by cattle and so make good cattle feed. The wood is also valued for carvings and furniture.

Propagation is by seeds which germinate during the rainy season, from the pods fallen on the ground.

FAMILY
Apocynaceae

NATIVE
Indian sub continent, China, Australia and Southeast Asia

FLOWERING TIME
July-August

SCIENTIFIC NAME:

Alstonia scholaris

COMMON NAMES :

Devil's tree, Scholar tree, Dita wocd, Blackboard tree (English);Saptaparani (Hindi); Aleale hale, Doddapapa (Kannada);Satvin (Marathi); Chattim (Bengali); Ezilaipillai (Tamil); Daevasurippi (Telugu); Daivappala (Malayalam)

An elegant tree, producing bunches of light green flowers, line the roads of many cities. With a pleasant fragrance pervading the surroundings, it is a pleasure walking under the tree when it is in bloom. The greenish-white or greyish -yellow flowers are borne in umbrella shaped bunches held in whorls of dark green shiny leaves.

The wood being weak is used for making pencils, blackboards and slates and so the common name 'Scholar tree'. It is the State tree of West Bengal and there is a tradition that The Chancellor of Visva-Bharati University, during its convocation, gives leaves of this tree to the graduating students. The tree also has many uses in traditional medicine.

The milky sap of the tree is poisonous, and probably the reason why people of certain communities, are reluctant to sit or even pass under the tree, believing that the devil resides in the tree and so the other common name of Devil's tree.

With medium height and a close set canopy the tree is most suited for avenue planting, both for its ornamental value and good shade. Old trees can be very tall and equally wide.

Propagation is easily done by the seeds that are produced in thin wiry pods which need to be collected in summer before they split open.

FAMILY
Fabaceae

NATIVE
Myanmar (Burma)

FLOWERING TIME
Feb to May

SCIENTIFIC NAME:

Amherstia nobilis

COMMON NAMES :

Pride of Burma, Tree of heaven (English), Urvasi (Hindi), Urvasi (Marathi), Shimshipavrisham (Malayalam) and Urbasi (Bengali)

These unusual looking crimson red, attractive flowers have earned the tree, the sobriquet of "Queen of flowering trees". Clusters of flowers hang down from a long stalk with each stalk containing 5- 6 flowers. Lasting for 2-3 days the flowers have five petals each of different sizes. The largest petal is fan shaped, with a wavy margin, white – yellow – orange in colour. Of the four crimson red petals, two are medium sized, with yellow highlights on the lips, and two are small. The stamens 9-10 in numbers are partially fused bearing prominent anthers.

This compound leaves are dark green in colour. When leaves are new they are flaccid pale tassels which turn purple before opening into green leaves.

The medium sized tree with a spreading canopy is suitable for avenue planting as well as for providing shade. It could make a beautiful specimen tree in landscapes of institutions and parks

The only method of propagation is by air layering. Although seeds are produced they do not germinate. This is probably the reason why this tree is not so popular.

Note : Few trees of this species have been recorded in Bangalore and Calcutta where it was first introduced in India.

FAMILY
Rubiaceae

NATIVE
South and Southeast Asia

FLOWERING TIME
June –August

SCIENTIFIC NAME:

Anthocephalus/ cadamba/ neolamarch

COMMON NAMES :

Bur flower (English); Haripriya (Sanskrit and Hindi); Kadamba (Kannada and Marathi); Kadam (Bengali); Katampu (Malayalam); Vellaikadabu (Tamil)

Unusual in shape and colour, the golden ball like structures, about 5cm in diameter bear clusters of yellowish white flowers looking like lighted bulbs hanging among the glossy green, elliptical leaves .With a strong fragrance, they attract bees, butterflies and birds.

In addition to the importance of various parts of the plant in ayurveda, the wood is useful in the plywood, paper & pulp industry. The flowers are distilled into the sandal wood to produce the traditional attar. The popular Apsara and Natraj pencils are made from this wood.

It is a tree often seen in temple complexes. With beautiful flowers and a broad crown, the tree is known becoming popular for planting on roadside and in parks.

The tree is held sacred by the followers of Krishna and its flowers are offered at the temples dedicated to him in memory of his swinging from the branches and dancing under the tree with the milk-maids of Vrindavan. It is said, that it was on this Kadamba tree that Krishna hung the clothes of the milk-maids, which he stole while they were bathing in the river Yamuna.

The ball like fruits when ripe contain a large number of trigonal and irregular seeds, which can be used for propagation. Cuttings and tissue culture plants are more common in recent times.

FAMILY
Meliaceae

NATIVE
Indian sub- continent

FLOWERING TIME
March-April

SCIENTIFIC NAME:
Azadirachta indica

COMMON NAMES :

Neem, Indian lilac (English); Neem (Hindi); Bevu (Kannada); Nimbay (Marathi); Maha neem (Assamese); Vembu (Tamil); Vepa (Telugu); Ayurveppu, Nimbam, Veppu (Malayalam)

When in full bloom, these star-shaped, white flower bunches hang under their own weight. Each bunch has 200-250 tiny flowers, where each flower is only about 5-6 mm long and 8-11 mm wide. When passing by a tree in full bloom, one can smell the flowers.

The long leaves bear serrated curved leaflets. The fresh green colour and shining surface of the leaves gives the tree a delicate and charming appearance.

Flowers and all other parts of the plant are known to possess one or the other medicinal property. Hence used right from cosmetic purposes to treating of cancer. Flowers are used for culinary purposes in some parts of the country while dried leaves are used to store grains & seeds and stem cuttings are used as tooth brushes. The tree has special significance in the celebration of the *Ugadi* festival in South India.

Flowers give way to oval fruits, which turn yellow when ripe. The seeds have gained much importance in the pesticide industry in the preparation of organic pesticides, soaps and bio diesel.

This evergreen tree with spreading branches is a common tree all over the country. It is planted along national and state highways, within city limits as well as on farm bunds. The tree with its spreading canopy makes a good shade tree and keeps the surroundings cool.

Propagation is by seeds and grafting.

FAMILY
Fabaceae

NATIVE
India

FLOWERING TIME
Feb to April & June to October

Bauhinia monandra

Bauhinia monandra

SCIENTIFIC NAME:

Bauhinia sps

COMMON NAMES :

Orchid tree, Butterfly tree, Camel foot (English); Lal kachnar, Kaniar, Kanchan (Hindi); Basavanapada, Banni (Kannada); Kurial, Kanchan (Assamese); Koiral (Bengali)

The Bauhinia genus of plants includes trees, shrubs and vines. Flowers resemble the orchid flowers and so they are known as Orchid Trees. The fragrant flowers are about 10-15 cm wide and are seen in a wide range of colours from red, purple, lavender, yellow to white and in various shades & streaks of the said colours. The size and shape of the petals also vary from narrow to broad and from straight to twisted. Flowers are borne in short sprays bearing two to three blooms

The most common of the Bauhinia flowers are *B.variegata, B.purpurea, B.acuminate and B. tomentosa*. All these Bauhinias have two things in common; one is the splash of colours of the flowers and the other is the rounded leaves split in the middle. The leaves resemble the hooves of ungulates and so the common name "Camel foot". Flowers are seen all round the year except for a short while in winter. They may be seen on leafless trees or along with the leaves.

The fruit is flat about ½ inch wide and 1 ft long with numerous seeds within. The young pods, leaves and buds are used in cooking in some communities. The trees yield a useful gum as well as oil from the seeds. The leaves and bark are used in tanning and dyeing. Apart from being grown for its ornamental value, the plant also has many medicinal properties.

The tree grows to a height of 20-40 ft with an almost equal size crown. Being a small to medium size tree it is most suited for growing along the city roads and medians.

Propagation is by seeds and semi-woody cuttings.

Bauhinia purpurea

Bauhinia tomentosa

Bauhinia acuminata

FAMILY
Malvaceae

NATIVE
India and Malaya

FLOWERING TIME
January – March

SCIENTIFIC NAME:

Bombax ceiba

COMMON NAMES :

Red silk cotton, Red cotton (English); Semal (Hindi); Burla – alli, Buruga (Kannada); Savar, Kate savar (Marathi); Shimul, Shalmali (Bengali); Sittan, Sanmali, Kongam (Tamil); Buruga (Telugu); Unnamurika (Malayalam)

After the leaves have fallen off in winter, the bare tree comes alive with the crimson red/orange flowers. Large flowers 14-16 cm wide are borne singly or in clusters. The cup shaped flowers hold water/nectar within. The flowers and the nectar are much relished by people of certain communities. These beautiful flowers are short lived, ending the crimson - red spectacle in about 15 days. They attract pollinating birds, squirrels and bees.

The flowers give way to green, fish shaped fruits which turn to brown on maturity. Once completely ripe the fruit bursts open to release soft silky cotton attached seeds which keep flying around. The silk cotton is also harvested and used for making pillows.

This is a fast growing tree used for its ornamental value. Every part of the tree is known to have medicinal value. It is a widely planted tree in parks, institutional gardens and along road sides

During evening times there is a cacophony of birds visiting the tree to savour the nectar from the flowers. Nature has it such that only birds can enjoy the nectar, as the trunk of the tree is spiked to prevent other animals from climbing up.

Propagation is from seeds and stumps.

White Bombax

FAMILY
Caesalpiniacea

NATIVE
Tropical America

FLOWERING TIME
February- March

SCIENTIFIC NAME:

Brownea coccinea

COMMON NAMES :

Scarlet flame bean, West Indian mountain rose, Rose of Venezuela (English); Lal Zumbar (Hindi); Lal Zumbar (Marathi); Supti (Bengal)

The scarlet flame bean tree is an evergreen, small tree with a well formed rounded canopy. The tree produces exotic looking, bright orange coloured flower heads, which hang upside down from the older branches. The flower heads are not easily visible on the surface and one can enjoy their beauty only by standing directly under the tree. Each flower head consists of 35-40 flowers with prominent yellow anthers peeping out.

Another interesting feature of this tree is the colour and texture of the young leaves. The pinkish-brown coloured leaves hang limply down from the tip of the branches. However on maturity they turn dark green and become stiff. Each leaf consists of 5 to 7 leaflets overlapping one another.

The plant can be propagated from seeds borne in dark brown coloured pods. Cuttings and air layerings also give good results. The air layered plants bloom in about 2-3 years while those grown from seeds may take upto five years.

It is not a common tree, and since it is amenable to so many means of propagation, the tree can be multiplied and popularised as an ornamental tree .

FAMILY
Fabaceae

NATIVE
Tropical and sub-tropical parts of the Indian Sub-continent

FLOWERING TIME
Jan to March

SCIENTIFIC NAME:

Butea monosperma

COMMON NAMES :

Flame-of-the-forest, Parrot tree, Bastard teak (English); Dhak, Palash, Tesu (Hindi); Muttuga (Kannada); Dhak, Palas (Marathi); Kincukam (Tamil);Kimsukamu, Palasamu (Telugu); Chamta, Kinsukam (Malayalam)

During the months of January- February when this crookedly shaped deciduous tree stands bereft of leaves, covered with the orange-red flowers, it is indeed a beautiful sight. Seeming surreal, it stands like an artist's creation. The scarlet red flowers appear as if the tree has been set ablaze. Initially the buds are black which burst into masses of orange flowers. Flowering is along the branches and at the ends. Each of the flowers has five petals, out of which one is larger than the others and which protects the other four. This large petal is in the centre and is pointed like a beak. When in bloom many birds visit the tree and so a good tree for bird watchers. It is the state tree of Uttara Pradesh.

The leaves are thick and velvety. In some parts of the country these leaves are stitched together with bamboo pins to make the ecofriendly, bio- degradable leaf plates and bowls. The fresh leaves are also used to sell minor forest fruits like *Karondas* and *Jamuns* during the summer and monsoon months. The dried flowers are used as safe colour for playing during *holi* (the festival of colours).

This medium sized tree being indigenous to India grows in almost all dry parts of the country. There are many traditional uses and rituals associated with the tree. So much has been written about the tree that it finds mention in both, the Ramayana and the Mahabharata. Leaves are trifoliate and considered to represent Vishnu, Brahma and Shiva.

The pods bear a single kidney shaped seed which is light in weight. The seeds are easily wind dispersed to germinate into new plants.

FAMILY
Fabaceae

NATIVE
Central and Northern parts of South America

FLOWERING TIME
March - May

SCIENTIFIC NAME:

Caesalpinia coriaria

COMMON NAMES :

Divi-Divi, American sumac (English); Sultan champa (Hindi); Vilayati alde kaayi (Kannada); Libi, Dibi (Marathi); Ingimaram, Inki maram, Konalkay (Tamil); b Divi divi, Divi divithumma (Telugu)

The clusters of greenish yellow flowers produced in the axils of the leaves, though not showy by themselves are perched amidst the feathery, fern like leaves. The flowers are interspersed with twisted pods which on ripening become black in colour.

The pods are rich in tannins and are used in the tanning industry. They are also used to produce blackish and blue dyes for colouring cloth and for painting. While the bark is used for preparing a red dye, the compound leaves are used for mulching.

This small tree with contorted trunk, with a round and spreading crown is mainly grown for its ornamental purposes as well as for providing shade. The tree can easily be shaped by pruning and training making it a good bonsai plant.

Propagation is by seeds.

FAMILY
Myrtaceae

NATIVE
Australia

FLOWERING TIME
In flushes all year round

SCIENTIFIC NAME:
Callistemon lanceolatus

COMMON NAMES :
Bottle brush (Eng); Cheel (Hindi)

These trees get their names from the spikes of flowers which bear a strong resemblance to a bottle brush. The obvious parts of the flower masses are stamens, with the pollen at the tip of the filament; the petals are inconspicuous. Flowers are arranged in a circular manner along the stems in the shape of a bottle brush. The filaments are usually red but white and pink colours are also known. Sometimes the pollen also adds a bright yellow to the flush of the flower spike. The bright red brushes of flowers provide a nice contrast to the dark green foliage. Leaves are narrow, pointed and arranged spirally along loose hanging stems. Each flower produces a small woody fruit containing hundreds of tiny seeds which open up only when the plant dies or fire causes the seeds to release. Pollinators, butterflies, and especially hummingbirds are attracted to this plant.

It is a fast growing plant with a hanging or weeping habit. It can grow into a tree upto 15 ft. or a shrub upto 6 ft. If pruned hard every year, the plant will grow in pots as small as 6 to 8 inches in diameter. The plant can be effectively used as a beautiful privacy screen hedge to block out unwanted views or as borders. It can be grown as an accent or specimen plant. It is a common tree for road median planting.

Propagation is by cuttings and air layering.

FAMILY
Clusiaceae

NATIVE
East Africa, Southern Coastal India, Malaysia and Australia

FLOWERING TIME
May - June

SCIENTIFIC NAME:

Calophyllum inophyllum

COMMON NAMES :

Alexandrian laurel, ball tree, Indian doomba oil tree, Beauty leaf, Leaf mahagony (English); Punnaga, Kamboja, Panchkesaru (Sanskrit); Surpun (Hindi); Huhonne (Kannada); Punnai (Tamil); Nameru, Ponna (Telugu)

The highly fragrant flowers of this tree are quite similar to that of *M suriga* . Borne in raceames of 4- 15 flowers, they are about an inch wide with a prominent set of stamens at the center. Used in worshipping and adorning the hair, the flower is considered sacred to Vishnu and is one of the flowers mentioned in the *Lalitha sahasrnama*. The beautiful leaves are leathery, oval, dark green and shiny.

It is a slow growing tree with a broad and irregular crown. Occasionally it grows in the wild, but is mostly cultivated in the coastal belts. Because of its attractive leaves, beautiful flowers and spreading crown, it has become a popular ornamental tree for planting along roadsides in cities and in public parks and gardens

The highly fragrant flowers have commercial value in the perfume, cosmetic & aroma industry as well as in Ayurveda. This tree is a potential species valued for its seed oil and is commercially being cultivated in South India for its oil. The tree has multiple uses: in agro-forestry, as a wind break and as soil stabilizer. The wood is popularly used in ship building.

New trees are grown from seeds as well as cuttings.

FAMILY
Caesalpiniaceae

NATIVE
Indian sub continent

FLOWERING TIME
March to April

SCIENTIFIC NAME:

Cassia fistula

COMMON NAMES :

Golden shower tree, Indian laburnum, Purging cassia (English); Arghavada (Sanskrit); Amaltas, Bendra lathi, Dhanbahev, Giri maloah (Hindi); Kakke mara (Kannada); Bahava (Marathi); Sonalu, Sonal, Amaltas (Bengali); Konrai (Tamil); Raela (Telugu); Kanikanna, Vishu konna (Malayalam)

This is one of the most beautiful tropical trees of Indian origin. The tree sheds its leaves at the end of the winter season and literally bursts into a mass of long, grape-bunch like yellow flowers. The bright yellow flowers are borne on bunches up to 2ft. long which hang like chandeliers. The short lived flowers are in bloom for about 15 days. These flowers are followed by green pods up to two feet in length which turn to black when mature. Each seed, encapsulated in a sweetish pulp is housed in a separated compartment within the foot-long pod.

The flowers are attractive to bees and butterflies and some species of butterflies choose the tender leaves of this tree to lay eggs on.

Since, parts of the tree are used for medicinal purposes, the Sanskrit name Arghavada means, a disease killer. The wood is hard and useful for furniture and inlay work.

To commemorate this beautiful tree a postal stamp has also been issued by the Indian Postal Department. It is the state flower of Kerala and the national flower of Thailand.

Propagation is from seeds, cuttings and air layering. However seed grown trees take as long as 8-10 years for blooming

FAMILY
Caesalpiniaceae

NATIVE
South America

FLOWERING TIME
March to April

SCIENTIFIC NAME:

Cassia moschata

COMMON NAMES :

Bronze shower tree (English)

This Cassiais is a show stopper with its orange -copper-yellow flowers. The drooping clusters of the colourful flowers cascade from the bare branches when the plant has shed all its leaves. However, some leaves start appearing along with the flowers.

Many wildlife visit the flowers for pollen and honey.

It is a fast growing tree reaching a height of 30-40 ft.Like most other Cassias this is a good ornamental tree suited for parks& gardens and plantingalong roads.

In its native the leaves are collected from the wild for medicinal purposes.

Propagation is from the seeds and half ripe wood cuttings

FAMILY
Fabaceae

NATIVE
Central and South America

FLOWERING TIME
February to April

Cassia javanica

SCIENTIFIC NAME:

Pink Cassia sps.

COMMON NAMES :

Pink shower tree, Coral shower tree, Stinking toe (English)

The lovely cool shades of the Pink Cassias, are a beautiful and restful sight. No one could fail to be moved by the glory of a Cassia in full bloom when its long, sweeping branches are clothed with the pink blossoms. They are one of the longest blooming flowers. Some start blooming as early as February and stay on the stem till August. The colour as well as size of the flowers varies with the species. Colour ranges from pink to purple which eventually fade to an orange.

After the end of the flowering season, green cylindrical pods are seen hanging down the tree. These turn brown with maturity, containing numerous seeds encased in a thick coloured liquid. This sticky liquid is known to be consumed by some communities.

The trees have a fairly wide and uniform canopy like an umbrella, thereby making a good shade trees. Because of the attractive flowers and shade providing nature, the tree is popular for avenue planting, in parks and institutional gardens.

Propagation is by seeds, cuttings and layering

CASSIA JAVANICA **:** Clusters of pink flowers are intermingled with the leaves

FAMILY
Fabaceae

NATIVE
Central and South America

FLOWERING TIME
February to April

Cassia grandis

***Cassia grandis*:** This Cassia is the first to bloom among all other Cassias. It blooms as early as February when the tree is bare sans the leaves. The flowers are smaller in size than other Cassias and are short lived

FAMILY
Fabaceae

NATIVE
Central and South America

FLOWERING TIME
February to April

Cassia renigera

Cassia renigera: It is a small tree compared to other Cassias. Except for a few leaves at the extremity, the branches are covered with a flow of white and pink flowers.

FAMILY
Fabaceae

NATIVE
Central and South America

FLOWERING TIME
February to April

Cassia marginata / roxburghii

Cassia marginata / roxburghii: also called the red Cassia. The short clusters of deep pink flowers hang from the sweeping branches. In April - May the tree seems weighed down with the flowers.

FAMILY
Fabaceae

NATIVE
South and Southeast Asia

FLOWERING TIME
June- January

SCIENTIFIC NAME:

Cassia simea

COMMON NAMES :

Yellow Cassia (English); Seemia, Kassod (Hindi); Simatangedu, Hiretangedi (Kannada); Manjal konrai, Manje konne (Tamil); Sima tangedu, Kurumbi (Telugu); Manja konna, Manja konnei (Malayalam)

Apart from the profuse blooming period from June-January, this tree produces one or two odd bunches all round the year. Yellow coloured pyramid shaped clusters of flowers are seen at the end of the branches. Each of the flowers is each about 3-5 cm long.

The seed pods about 25cm in length are soft green and ribbon like which turn brown when ripe. These dry pods are seen hanging in dense clusters, giving the tree an unkempt look.

This small to medium sized evergreen tree has multiple uses in agroforestry. The leaves are used as green manure and a row of well pruned trees can make a good live fence. The spreading canopy provides shade in cocoa, coffee and tea plantations. It is suitable for planting along roadsides in cities and on highways.

The wood of the tree is used in construction, furniture making and cabinet & inlay work. The heart wood and fruits are useful in traditional medicine.

Propagation is by seeds, stumps and nursery grown seedlings.

FAMILY
Fabaceae

NATIVE
Australia

FLOWERING TIME
February –March

SCIENTIFIC NAME:

Catanospermum australe

COMMON NAMES :

Black bean tree, Australian chestnut (English)

This is a large evergreen tree with a dense crown of dark green glossy leaves. In summer sprays of yellow-orange-red flowers are seen blooming directly on the trunk and stem of the trees. The flowers are frequented by birds, bees, butterflies for their nectar.

The dark brown seed pods look equally attractive on the plant. The timber of the tree is strong and durable and so useful for carving and cabinet making.

It is ideal for growing as a shade tree in parks, gardens and roadsides. The root system is known for fixing nitrogen.

Propagation is from seeds.

FAMILY
Verbenaceae

NATIVE
Tropical American

FLOWERING TIME
June –July

SCIENTIFIC NAME:

Citharexylum spinosum

COMMON NAMES :

Fiddle wood tree (English); Manjari (Hindi); Beli Parijatha (Kannada); Parijatham (Malayalam)

Flowers of this small evergreen tree are small (0.3inches) and richly scented. The tassel shaped inflorescence on which they are borne hang down the tip of the branches. When these little glowers fall to the earth it is indeed a beautiful sight. The dark green oval shaped leaves set the backdrop for the pure white flowers. The tree could be helpful to honey collectors as the flowers are foraged by bees for the nectar.

Trees grown in the wild are harvested for local use as food, medicine and good quality of wood. Since the wood is popularly used for making musical instruments, hence the common name of Fiddle wood tree.

A fast growing tree grown for its ornamental value, makes a good roadside tree.

Propagation is by bird dispersed seed or suckers from coppiced trees.

FAMILY
Fabaceae

NATIVE
Madagascar

FLOWERING TIME
February -March

SCIENTIFIC NAME:

Colvillea racemosa

COMMON NAMES :

Colville's glory, Whip tree (English); Kilbili (Hindi) Hukuchhina mara (Kannada)

A highly decorative plant that produces spectacular cone shaped clusters of bright red-orange flowers. The grape - like clusters of velvety buds, varying from green to blood orange, are a sight to behold. The cylindrical bunches of flowers about 1-1 1/2ft, comprise hundreds of flowers. These hang down the tip of the branches. The yellow-orange stamens are the conspicuous part of the flower and attract bees, butterflies and birds. Since the blooming season is quite short, one must make it a point to visit this plant to savour the beauty.

The foliage is evergreen. The leaves are fern like, light and feathery looking, very much similar to that of the Gulmohar tree. This fast growing upright tree gains a height of upto 50 ft. as it ages. Although it requires similar growing conditions of that of the Gulmohar (*Delonix regia*) tree, this plant has not gained the attention of the city planners and is not so common. Being a medium to large sized tree, it could make a good roadside tree and for plantings in large gardens & parks where space is not a constraint.

All parts of plant are poisonous if ingested. Handling the plant may cause skin irritation or allergic reactions. Plant has spines or sharp edges so caution may be exercised when handling.

New trees are grown from seeds. The mature dried seeds can be collected on the plant by bagging. For germination the seeds need to be scarified.

FAMILY
Boranginaceae

NATIVE
Southern USA to
Central Mexico

FLOWERING TIME
All year round

SCIENTIFIC NAME:
Cordia boissieri

COMMON NAMES :
Mexican olive, White cordia

The white trumpet shaped flowers with yellow throats put up a show all year round. The fragrant flowers are each about 1.2 -2 inches wide.. The thin and sheer petals can be described as looking like crepe paper or chiffon.

The silvery green leaves are velvety textured. The olive-like, white fruits that follow the flowers have a sweet flesh relished by birds and other wildlife.

This small, evergreen tree grows to a height of only 15-20 ft. with a spreading canopy. The compact growth and showy flowers all round the year make it a popular ornamental tree suitable for placing in a lawn area as a free-standing specimen. It can be also be planted in a container and with good maintenance can be kept looking nice for a number of years.

The wood is used as firewood and in carpentry. Although fruits are edible the quantity of consumption may be limited.

Propagation is by seeds and air-layering

FAMILY
Boraginaceae

NATIVE
North and Central America

FLOWERING TIME
All round the year

SCIENTIFIC NAME:

Cordia sebestena

COMMON NAMES :

Broadleaf geiger, Scarlet cordial (English); Lal lasoda, Bohari (Hindi); Kendachalle-gida, Challe-kendala (Kannada); Kamla buhal, Raktarag (Bengali); Acchinaruvihil (Tamil); Virigi (Telugu)

Flowering almost all year round, the clusters of orange coloured flowers are produced at the end of the branches. Heavy flowering however, is during spring and summer. Each orange coloured flower is tabular with a width of two inches. The flowers are followed by pear shaped fruits which are edible.

It is an ornamental tree grown for its evergreen leaves and colourful flowers. It is suitable for growing in gardens and along road sides. The rounded crown makes it a good shade providing tree.

Propagation is by cuttings and layering. Seed sown trees are slow to germinate and take a longer time to bear flowers.

FAMILY
Lecythidaceae

NATIVE
Tropical South America and West Indies

FLOWERING TIME
Almost all round the year

SCIENTIFIC NAME:

Couroupita guianensis

COMMON NAMES :

Cannon ball tree, Christ in the cradle, Monkey pots (English); Nagaling, Tope gola (Hindi); Lingada-mara, Nagalinga pushpa (Kannada); Nagkesar (Bengali);Nagalingam, Lingam (Tamil); Nagamalli (Telugu)

This large tree indigenous to the Amazon rainforest is considered sacred in India because of the uncanny resemblance of the flower to the *Shivalinga* guarded by the many headed cobra. The *Sivalinga* shape is visible at the center of the flower while the pollens are on a structure which is shaped like a serpent hood. The flowers are borne on thick tangled extrusions on the lower part of the trunk of the tree just below the foliage. These attractive unusually shaped flowers have a strange but sweet perfume, particularly strong during night time. The six petals are bright pink to red, or can be multicoloured with white and yellow on them. Bees, butterflies and birds love these flowers. It is the state flower of Pondicherry.

On the trunk and main branches are borne the large globosely shaped woody fruits which look like big rusty cannon balls hanging from strings. The fruits contain small seeds in a white, unpleasant smelling jelly. The hard shells of the fruit are shaped for use as containers and utensils.

This is an evergreen tree with the various parts being used for medicinal purposes. The wood has little or no commercial value but is cultivated for the attractive flowers along road sides, in temple complexes and in parks. Although literature cites that this tree is rare in India, it is quite common in South India and more specifically Karnataka.

Due to danger of falling fruits, the planting of this tree should be avoided in vehicle parking spaces and along paths where people pass by.

Propagation is by seeds.

FAMILY
Capparaceae

NATIVE
Tropical Asia

FLOWERING TIME
March - May

SCIENTIFIC NAME:

Crateva religiosa

COMMON NAMES :

Barma tree, Spider tree, Temple plant, Sacred Garlic Pear (Eng); Barna, Vaivarno (Hindi); Holenekki (Kannada); Chinnaulimidi, (Telugu); Adhicharam (Tamil) Killi, Neervala, (Malayalam); Tikta-shaka, Varuna (Sanskrit)

The common name of Spider tree is because of the long spidery stamens of these beautiful flowers. The cream/white petals are flimsy and almost transparent. A variety of fauna visit the nectar filled flowers. The trifoliate glossy green foliage provides a good backdrop to the white /yellow flowers..

It is a slow growing, much branched small tree. Generally an evergreen tree but can become deciduous where winters are severe. Suited for growing in large residential gardens, in parks and along road sides. Considered a sacred tree, it is often grown in temple complexes and near graves.

The various parts of the plant are used for medicinal purposes. The wood is used locally in villages to make drums and artefacts.

Propagation is by seeds, semi hardwood cuttings and suckers

FAMILY
Fabaceae

NATIVE
India, Sri Lanka, Nepal, Burma

FLOWERING TIME
April -May

SCIENTIFIC NAME:

Dalbergia lanceolaria

COMMON NAMES :

Bastard Rosewood (English); Takoli (Hindi); Bili bete (Kannada); Dandus (Marathi); Erigai (Tamil); Nagulapachari (Telugu)

This profusely blooming tree is a handsome one with tiny flowers completely covering the entire canopy when in full bloom. The small pinkish-white flowers appear in bunches at the end of the branches.

Although the tree is grown for its good quality of wood the flowers when they last for about a month are worth enjoying.

A medium sized tree, it is a good avenue tree and for planting along field borders.

The flowers are followed by thin, flat pods tapering at both ends. These longish pods hold a single seed within.

The tree has many medicinal properties. While the bark is used as an analgesic and anti-diarrhoeal, the sweet blackish pulp of the seed pod is used as a mild laxative.

Propagation is by direct sowing of seeds.

FAMILY
Fabaceae

NATIVE

India, Nepal and Indonesia

FLOWERING TIME
July and August

SCIENTIFIC NAME:

Dalbergia latifolia

COMMON NAMES :

Indian rose wood, Black rose wood (English), Kala Shisham (Hindi) and Beete (Kannada)

The tree grown for its timber produces small white flowers in inflorescences measuring about 5 to 20 cm long. The heavily laden inflorescence amidst the small rounded leaves is a beautiful sight, although the period of flowering is short lived. When in full bloom, the tiny little flowers keep falling to the ground all day long carpeting the area below. The flowers give way to brown pods about 5 cms long, pointed at both the ends.

The wood of this tree is used for fine furniture and all kinds of construction work. It is the most popular wood for carving and engraving.

It is a single stemmed tree with a dome shaped crown. The round canopy makes it a good shade tree for planting along road sides and in parks & gardens. It is also a very important tree in agro-forestry

The common forms of propagation are by seeds, roots and stem cuttings.

FAMILY
Fabaceae

NATIVE
India

FLOWERING TIME
All round the year

SCIENTIFIC NAME:

Delonix elata

COMMON NAMES :

White gulmohur, Creamy peacock flower (English); Waykaran, Samrsro (Hindi); Kempukenjiga, Gentige hoo (Kannada)

Being a species of *Delonix* the tree is similar to the red Gulmohar, except for the colour of the flowers. Initially they are white, gradually turning to yellow. The upper petal is smaller and darker in colour, almost circular, with wavy edges. While the red Gulmohar blooms profusely for a short while during the month of May, this one has flowers all round the year.

The foliage and young leaves are same as that of *D. regia* and are used as fodder for goat, sheep, camel and cattle. It is widely grown as an ornamental or as a shade tree alongside roads. It is useful in farm forestry to control soil erosion and for reforestation of difficult sites.

It is a fast growing tree raised easily from seed.

FAMILY
Fabacae

NATIVE
Madagascar

FLOWERING TIME
April –June

SCIENTIFIC NAME:

Delonix regia

COMMON NAMES :

Royal poincciana, May flower (English); Gulmohar (Hindi); Kempu torai, Kattikai mara (Kannada); Krishnachura (Bengali); Mayil kondrai (Tamil)

True to its name the "May Flower" rarely fails to keep its promise to the month of May. This is probably one of the most spectacular of the tropical flowering trees. During the peak of summer it bursts open into orange, bright red flowers. Each individual flower has five petals, one of which is slightly larger than the others having white / yellow spots. The red canopy lasts for about a month. Most often the tree is sans the leaves when in bloom, but the delicate & feathery dark green pinnate leaves sometimes appear along with the flowers.

Fruits are long bean- like flat pods, which initially are green in colour, which turn to dark brown on maturity. Dangling from the bare branches the fruits stay on the tree almost till the next flowering season or may be seen along with the flowers. Unless of course they are harvested when green for feeding the cattle or for fuel purpose when dry.

Mostly grown for the beautiful flowers, the white, soft wood can be polished and used for making ornaments. Some people use the flowers and buds for flavouring foods.

Being a very popular tree in cities and country sides children of yester -years enjoyed playing with the various parts of the plant. The brown pods became swords for battling, the red calyx became nail extensions, the white petals were eaten for their sour-sweet taste and the tiny leaflets became confetti for showering on each other.

This ornamental tree with attractive flowers & dense foliage and with a spreading canopy makes a good shade tree.

Propagation is from seeds.

FAMILY
Malvaceae

NATIVE
Madagascar

FLOWERING TIME
November –January

SCIENTIFIC NAME:

Dombeya spectabilis

COMMON NAMES :

Dombey, Maple leaved dombey (English); Domrupani (Bengali)

This is a small tree, sometimes considered a large shrub which produces rounded heads of pinkish –white flowers. Each of the flower head consisting of numerous flowers, literally hang down under their own weight. In the early mornings one can notice heavy activity of the bees collecting pollen and honey from the flowers.

Leaves are large and maple leaf shaped. In villages and the countryside this plant grows wild yielding fibre from the bark for rope making, spinning and weaving.

The plant is easy to grow and can make a good ornamental plant in residential gardens as a standalone specimen plant or a hedge plant. It could be effectively used on median strips of highways.

Propagation is by seeds, semi hard wood cuttings and air layering.

FAMILY
Fabaceae

NATIVE
South America

FLOWERING TIME
February-April

SCIENTIFIC NAME:

Erytrina mitis

COMMON NAMES :

Shady Coral tree (English)

The Erythrina is a large genus with over 100 species of plants mostly from the tropical and sub tropical areas of the world. Though they are broadly termed as coral flowers, the colour is more often red than that of corals. Within the species the plant sizes also vary from giant size trees to small shrubs with varying size of flowers, shapes and inflorescence. Erythrina trees are used as living fences, wind breaks for soil and water conservation, shade tree for coffee plantations, support trees for pepper vines, and for the timber. The flowers of the tree are eaten or used to make tea and the seeds are used to make jewellery. The leaves are used as animal fodder and the wood can be carved into canoes and surf boats.

Erythrinas are easy to grow drought tolerant plants mostly grown for the attractive flowers. They can be grown from seeds, hard wood cuttings and air layered cuttings

Presented here are Erythrina trees which are commonly seen in landscapes of large gardens of homes and institutions and in parks.

Erytrina mitis is a medium size tree grown for shade purposes, produces masses of orange-yellow flowers.

FAMILY
Fabaceae

NATIVE
India

FLOWERING TIME
Feb-April

SCIENTIFIC NAME:
Erythrina indica

COMMON NAMES :
Indian Coral tree (Eng); Pangara (Marathi); Pangari, Dhadak (Hindi); Kalyana murugai (Tamil)

Like other deciduous flowering trees, this tree is also bereft of leaves when it blooms. At this time the entire canopy is covered with clusters of scarlet coloured flowers. The colourful and attractive flowers make this tree a highly valued ornamental tree, which has therefore been described as one of the gems of the floral world. There is also a variety with white coloured flowers. At the end of the flowering the compound leaves with three leaflets start appearing.

The nectar filled flowers attract a large number of birds like crows, mynas and parrots. These birds feed on the nectar and aid in pollination. The new branches have prickles which work to the advantage of the birds in savouring the nectar and preventing other animals from the same.

This is a fast growing, drought resistant medium sized tree. It is grown as an ornamental tree, a living fence, hedge plant, medicinal plant, shade tree and for soil conservation Since it is a good nitrogen fixing plant, farmers plant it in their fields to increase yields and as a support tree for climbing plants like betel vines, black pepper, . The wood is light and soft and so is used where hard woods cannot be used

Seeds harvested within 3 - 6 months of maturity, can be sown to grow new plants

FAMILY
Fabaceae

NATIVE
India

FLOWERING TIME
Feb-April

SCIENTIFIC NAME:

Erythrina blakei

COMMON NAMES :

Blake's coral Tree (English) and Halvana (Kannada)

Deep scarlet flowers are borne in clusters at the end of branches. The elongated bunches can be upto 3 ft long. Unlike some species this species flowers when the tree is still covered with leaves.

This small tree attracts a lot of birds.

FAMILY
Fabaceae

NATIVE
India

FLOWERING TIME
January to March

SCIENTIFIC NAME:

Erythrina crista – galli

COMMON NAMES :

Cockspur coral tree, Cry – baby tree (English)

Flowers are borne in long clusters, which sometimes are several feet in length and form a part of the branch. Individual flowers are single beak shaped, red to pink in colour.

It is a densely branched tree, that can be pruned to keep it low in height to suit home gardens.

FAMILY
Sapindaceae

NATIVE
India (Western ghats), Sri Lanka and Africa

FLOWERING TIME
March- May

SCIENTIFIC NAME:

Filicium decipiens

COMMON NAMES :

Fern tree, soap berry (English); Kotu, Puveras (Hindi); Neeroli, Kaadre hoovarasi (Kannada); Ningal (Tamil); Pattakunkudu (Telugu) Val murichcha (Malayalam)

This plant is better known for its foliage than the flowers because flowers last only for a short time. The small flowers which are white with a tint of pink are produced in large clusters (6-8 inches long). The loose clusters are erect and upright but most times are hidden by the dense foliage.

The large fern like leaves of the tree are attractive both because of the glossy bright green colour and the neat arrangement of the leaflets along the prominent midrib. They are evergreen and remain on the tree throughout the year. The attractive green colour, and their ability to stand straight and upright make it suitable as cut foliage in bouquets and arrangements.

It is a slow growing evergreen tree cultivated in gardens and landscapes for its lush green, attractive foliage, compact size and the uniform crown. The round crown makes it a good shade tree for avenue planting. The small neat look makes it ideal for residential garden.

The wood is hard and used in heavy construction. Being a slow burner the wood also makes good fire wood.

New trees can be grown from seeds which should be sown quickly within few days of harvesting. Birds eat the fruits and disperse the seeds.

FAMILY
Fabaceae

NATIVE
Tropical dry forest in Mexico, Central America and Central parts of India

FLOWERING TIME
January – February

SCIENTIFIC NAME:

Gliricidia sepium

COMMON NAMES :

Gliricidia, Mexican lilac (English); Gobbaragida (Kannada); Giripushpa (Marathi); Saranga (Bengali); Siima konna (Malayalam)

This tree though not really grown for the flowers, is well worth its beauty when in bloom. During the months of January- February the bare stems of the plant, having shed all their leaves are wrapped with the lilac coloured flowers. Each small, pink-lilac flower with a white tinge usually has a pale yellow spot at the base.

This is a small to medium size, fast growing tree considered as nature's best gift for organic farming. The compost prepared from the leaves and twigs as well as fresh leaves for green manure are good substitutes for chemical fertilizers.

It is a multipurpose tree used for timber, fire wood, medicines and for live fences. Apart from acting as a shade tree in plantations, it prevents soil erosion and stabilizes the soil . The tree is ideal for planting along farm bunds so as to serve the dual purpose of green leaf manure as well as to enjoy the beauty of the pleasant pink flowers.

Gliricidia can be propagated from seeds as well as through stem cuttings.

FAMILY
Lamiaceae

NATIVE
Asia

FLOWERING TIME
February to April

SCIENTIFIC NAME:

Gmelina arborea

COMMON NAMES :

White Teak (English); Gambhar, Bhadrapani (Hindi) Shivani (Kannada); Shivan, Shewan (Marathi); Gambhar, Gamar (Bengali); Kumishan, Kumizhan (Tamil)

Flowering occurs when the tree is almost leafless. The yellowish- brown flowers easily merge with the colour of the bare branches making the flowers inconspicuous and may go un-noticed. They occur at the tip of the branches in narrow branching clusters. Each flower is 1-2 inches wide with five lobes. These lobes are in the form of a gaping mouth, resembling the snap dragons.

It is a medium to large sized deciduous tree with very good quality timber. The trees are harvested from the wild for local use as food, medicine and source material. The wide spreading canopy renders it suitable for ornamental planting and for providing shade along roadsides or in gardens and parks.

Propagation is by direct sowing of seeds or transplantation of seedlings.

FAMILY
Proteacea

NATIVE
Australia

FLOWERING TIME
March – April

SCIENTIFIC NAME:

Grevillea robusta

COMMON NAMES :

Silver oak, Silky oak (English); Chappari-ro-paid (Hindi); Naati oak mara (Kannada)

Being a very tall tree one needs to really strain the neck to spy the unusual flowers. Often they go unnoticed because of the height of the tree. The golden yellow tube like flowers, appear in clusters on long spikes, resembling the scrubbing brushes. A closer look at the flowers will show that each tiny bloom within the brush-like structure is curved.

The fern like olive green leaves are shining on the upper surface with silvery colour on the undersides. This silvery surface is momentarily seen when the wind blows.

It is a fast growing, evergreen, tall tree reaching a height of even 90-100 ft. It is grown for ornamental purposes as well as a wind breaker and a shade tree in tea and coffee plantations. It can make a good boundary marker tree.

The tough and long lasting wood fetches a good price in the market and so growing silver oak commercially for the wood is also catching up. The wood is used in carving, panelling, furniture making, for tennis racquets as well as pulped to produce paper.

Flowers rich in nectar are attractive to bees, butterflies and birds. Honey cultivation could be an enterprise where many oak trees are planted.

However all parts of the plant are considered poisonous if ingested. Even contact with the plant may cause skin irritation or allergic reactions.

Seeds germinate easily when sown during the monsoons.

FAMILY
Lecythidaceae

NATIVE
Central and Northern South America

FLOWERING TIME
All year round

SCIENTIFIC NAME:

Gustavia superba

COMMON NAMES :

Membrillo, Sachamango, Heaven Lotus, Monkey fruits (English)

The pink - purple flowers with waxy petals are borne directly on the stem. They are fairly large having a diameter of 4-6 inches with a ring of yellow stamens at the center. Blooming almost all round the year there are a few flowers always dotting the tree. The lanceolate leaves are a bright glossy green

It is a small evergreen tree growing about 40 -50 ft. and is valued for its ornamental properties. Although not a common tree outside its nativity, it grows well in tropical countries and could be used in countries for planting along roadsides. The fine grained strong wood is used to some extent in construction. In its native it is cultivated in home gardens for the fruits.

Fruits are globose with a few seeds held within. They are edible and are supposedly rich in Vitamin C. They are consumed raw or added to soups, rice and meat.

Propagation is by seeds.

FAMILY
Bignoniaceae

NATIVE
South America

FLOWERING TIME
April onwards till August

SCIENTIFIC NAME:

Jacaranda mimosifolia

COMMON NAMES :

Blue jacaranda, Fern tree (English); Neeli gulmohar (Hindi); Neeli turai (Kannada)

The Jacaranda is a beautiful tree bearing attractive clusters of purple to deep blue-mauve colour flowers. Bunches of the flowers can be upto 1 ft long, while each tubular flower is about 2 inches long and 1½ inches wide. The striking blooms can appear anytime from April to August. The effect is more spectacular in colder regions, when all the leaves are shed and the tree is swathed in blue flowers. The withering flowers cover the ground beneath the tree with a carpet of blue flowers.

The bi- pinnate compound leaves with a pleasant green colour make the tree attractive even when not in bloom. The leaflets are delicate and look like ferns and so the tree is also known as "Fern tree". After the flowers fall off, flat and round shaped seed pods are formed.

The tree has no commercial value except for ornamentation. It is a fast growing tree attaining a height of upto 25-40 ft. It is most suited for planting along roadsides, in parks and institutional gardens. If space is not a constraint the tree could be grown in residential gardens also.

New trees can be grown from seeds.

FAMILY
Bignoniaceae

NATIVE
Mozambique and parts of Tropical Africa

FLOWERING TIME
May to July

SCIENTIFIC NAME:

Kigelia africana

COMMON NAMES :

Sausage Tree (English); Balam khira, Jhar fanoos (Hindi); Aanethoradu kaayi, Mara sowthae (Kannada)

This tree produces unusual flowers and fruits hanging on rope like structures. Each dangling stalk has numerous flowers and each individual bell shaped flower is blood red in colour with a velvety texture inside. However the individual flower is horizontally oriented to the stalk. The fragrant, nectar-rich blossoms are pollinated by bats, insects and sunbirds. The flowers bloom by night and by morning one can notice the flowers strewn on the ground below, while still looking fresh.

The other interesting feature of this tree is the mature woody fruits which are up to two feet long. They look like giant cucumbers or sausages, dangling from the long stalks.

In its native-Africa, the tree has many traditional medicinal uses. There are some steroids/chemicals found in the sausage tree that are currently added to commercially available shampoos and facial creams. The wood is hard and used for cabinet making.

The medium to large tree grows in drier parts of the country. It has no commercial value and is mainly grown for the curious looking, ornamental flowers and fruits. Being an evergreen tree with a shapely crown, it is ideal for parks, institutional gardens and roadside planting.

Seeds are extracted by pounding the dry pods. The sooner they are sown after collection better is the germination as the viability of the seeds is only for a few days after harvesting.

FAMILY

NATIVE
Tropical Southern Asia

FLOWERING TIME
March-May

SCIENTIFIC NAME:

Lagerstroemia flos reginae / speciosa

COMMON NAMES :

Crepe Myrtle, Pride of India (English); Jarul (Hindi); Taman (Marathi); Kadali (Tamil)

Lagerstroemia is a beautiful summer flowering tree producing spikes of pleasantly coloured flowers in pink, lavender and pink-mauve. Each individual flower on the spike is about 2" wide. The flower petals are crumpled and wrinkled and hence the common name of crepe flower. It is the state flower of Maharashtra.

The flower spikes later develop into large woody capsules bearing seeds with apical wings. These are easily dispersed by wind. The seed bunches are not very attractive when left on the tree but could be effectively used in dry flower arrangements.

It is a small to medium sized deciduous tree shedding the leaves gradually, so the tree is never completely bare.

The tree is supposed to have many medicinal uses. The wood which is strong and termite resistant is second only in quality to teak and is used for house construction and boat making. The tree however is more popular for its ornamental value than for the wood.

The tree is suitable for planting along roadsides in parks & institution gardens. Being a small tree it could also be planted in large residential gardens.

Propagation is from soft wood or semi-ripe cuttings. The tree comes to bloom within 2-3 years of planting.

FAMILY
Magnoliaceae

NATIVE
Indo Malaya region

FLOWERING TIME
March - August

Red champaca

SCIENTIFIC NAME:

Magnolia champaca

COMMON NAMES :

Champak (English); Champa (Hindi); Sampige (Kannada); Pivalu champa (Marathi); Champa (Bengali)

The champaka flowers are considered sacred for worshipping by Hindus and Bhuddists. They are normally seen in three colours namely: white, yellow and orange. The fragrance of each of the three varieties is quite distinct. The white smells the strongest, followed by the yellow and the mildest of them is the orange coloured. It is this fragrance, that has made these flowers very popular.

The leaves are ovate, fairly large, bright green in colour. Being semi deciduous the leaves are shed gradually over the winter. The tree is therefore never completely bare.

This tall tree as it grows older may reach a height of up to 90-100 ft. The tree produces wood of fine grain, suitable for building construction, furniture, toy making, and carving. Essential oil extracted from the flowers is used in the perfume industry. The tree is seen in large home gardens and temple complexes. It is now popularly grown as an avenue tree along roadsides of residential layouts.

Trees propagated from seeds take 8-10 years for blooming, whereas if grown from cutting and layers can flower in 2-3 years.

Yellow champaca

White champaca

FAMILY
Magnoliaceae

NATIVE
Indo Malaya region

FLOWERING TIME
June - August and sporadically

SCIENTIFIC NAME:

Magnolia coco

COMMON NAMES :

Champak (English); Champa (Hindi); Sampige (Kannada); Pivalu Champa (Marathi); Champa (Bengali)

The flowers of this tree bloom in the evening with a fruity smell. The fragrance is at its best in the morning. The flowers borne at the end of the branches open up in the evening and before these flowers actually open, the buds look like hanging eggs enclosed in green sepals. Once open, they last only for a day to spread their beauty and fragrance. The dark green smooth and leathery leaves present a pleasing background to the cream/ white flowers.

Being small in size, this evergreen tree is sometimes classified under shrubs. Growing to a height of about 6-8 ft. the tree could be planted as a hedge plant or as a standalone tree. It is also amenable for growing in pots where it stays at a height of 2-3 ft.

Propagation is by cutting and layers

FAMILY
Cluiaceae

NATIVE
Western Ghats of India

FLOWERING TIME
March to July

SCIENTIFIC NAME:

Mammea suriga

COMMON NAMES :

Naga –Kesaram (Sanskrit); Surangi (Hindi, Marathi, Konkani);Surgi (Kannada); Nagapu (Tamil); Surapunna (Telugu, Malyalam); Nageswar (Oriya)

This medium sized tree, bearing fragrant white flowers is endemic to the western ghats. The beautiful rounded white/pinkish buds bloom into four petalled white flowers. The petals are round and wide with a mass of yellow stamens at the center. The fragrance of the flowers is such that the smell permeates upto a radius of 500 metres.

The fresh flowers are used for worshipping and adorning the hair. Essential oils are extracted from the dried flowers for use in the cosmetic and perfume industry. The silky red colour extracted from the dry flowers is used as a natural dye in the textile industry.

The tree is suitable for avenue planting and as a household plant

The tree produces a single seeded ovoid fruit. Propagation is by seeds.

FAMILY
Anacardiaceae

NATIVE
South Asia

FLOWERING TIME
Dec -January Feb

SCIENTIFIC NAME:

Mangifera indica

COMMON NAMES :

Mango (English); Aam (Hindi); Mavina mara (Kannada); Amba (Marathi); Ma (Tamil); Mamidi (Telugu); Mangga (Malayalam)

The word mango conjures the picture of the luscious mango fruit: the national fruit, the king of fruits and the most revered fruit. Though not grown for the flowers, the mango inflorescence needs no less appreciation. Heavily laden trees in the months of December-January present a breath-taking sight. Each inflorescence with hundreds of tiny lemon yellow to red florets emit a musky smell spreading the heady fragrance in the surrounding areas. Being an alternate bearing plant there are on-years and off-years of blooming. However in recent times newer varieties bloom every year to produce fruits.

With the blossoming of the trees, begin the common man's speculation about the yield of the fruit, arrival in the market and the price thereof. Although flowers bloom in plenty, less than one per cent (1%) of them mature into fruits.

Propagation is by seeds. New varieties are often grafted on local root stock.

FAMILY
Bignonaceae

NATIVE
Eastern Africa

FLOWERING TIME
All round the year

SCIENTIFIC NAME:

Markhamia lutea

COMMON NAMES :

Nile Tulip (English)

The bright yellow- orange coloured flowers of this small tree look beautiful with the back-drop of the large, dark green leaves. The flowers are borne in clusters at the tip of the branches. Each trumpet shaped flower with frilly lobes is about 6 cm long.

The wood of the tree is fairly resistant to termites and therefore very useful in making furniture, hand tools, boats etc. The leaves are used as mulch for moisture conservation and increasing organic content of the soil.

Being a small tree it is suitable for planting as a screen or background tree for large gardens. Since it does not branch much, it is suitable for growing along narrow lanes where space is a constraint.

After flowering thin, long and twisted brown seed pods are seen hanging on the tree. When the pods open in summer, the seeds are dispersed by wind. These seeds soon sprout during the rainy season. Many seedlings can be noticed in the vicinity of a mature tree.

Flowers provide good forage material for the bees.

FAMILY
Meliaceae

NATIVE
South Asia

FLOWERING TIME
February - May

SCIENTIFIC NAME:
Melia dubia / azedarach

COMMON NAMES :

Indian lilac, Bead tree, China berry tree (English); Bakain (Hindi); Hebbevu (Kannada)

In spring and early summer, the Indian lilac produces drooping bunches of purplish flowers. The bunches consist of fragrant, star shaped flowers, each about 3/4 " in diameter.

This tree could easily be confused with Neem. However, there is a variation in the structure of the leaves and the colour of the flowers. The Neem leaves are smaller in size and curved whereas, Melia are larger and straight. Though the structure of the flowers is exactly the same, Neem flowers are white and those of Melia are lavender. While Neem is not considered as poisonous, Melia is considered to be poisonous.

Flowers are followed by clusters of spherical, yellow fruits about three fourth inch ($3/4^{th}$"), in diameter that persist on the trees even after the leaves have fallen. All parts of Persian lilac tree are poisonous.

In recent times, it is gaining importance as an agro-forestry tree because of its use in bio-fuel production and in the plywood & paper industry. The fast growing and hardy nature of the tree has made it important in reforestation and regeneration of waste lands. It also makes a good avenue tree and is seen planted in many city landscaoes.

Propagation is by seeds.

FAMILY
Callophyllaceae

NATIVE
Tropical parts of Sri Lanka, India and Malay peninsula

FLOWERING TIME
Winter through Summer

SCIENTIFIC NAME:

Mesua ferrea

COMMON NAMES :

Indian iron wood, Cobra Saffron (English); Nag Champa, Nagkesar, (Hindi); Tadinangu (Tamil); Thorlachampa (Marathi):Vainavu (Malayalam) Nokte (Assamese); Nageshor (Manipuri)

The large white flowers with orange -yellow stamens at the center possess a single style that resembles the hood of a cobra. The large, 2-3 inches wide, fragrant flowers attract a lot of bees, birds and butterflies. Flowers open early morning and last only for a day. They are considered auspicious and used for worshipping.

After blooming in the dry months, young leaves with a pinkish red flush begin to appear at the onset of the monsoons.

Growing in the Western Ghats of South India it is a tree of the evergreen forests, especially the river valleys. This tree with a weeping habit grows up to 60 ft tall, but is a slow growing one. It is national tree of Sri Lanka, state tree of Mizoram and state flower of Tripura.

This is a widely cultivated ornamental tree seen near temples and is well suited for large gardens/ parks, farm houses and for planting along roadsides.

The wood of the tree is as hard and heavy as iron and so the species name of ferrea. This property makes it hard to work with the wood. Various parts of the tree are used in Ayurveda, Siddha and Unani systems of Indian medicines for treating many diseases.

Fruits are normally green in colour, changing to brown as they mature. One fruit contains 1-4 seeds.

Propagation is by seeds only because cuttings will not produce the tap root.

FAMILY
Fabaceae

NATIVE
Myanmar and Thailand

FLOWERING TIME
February – March

SCIENTIFIC NAME:

Milletia ovalifolia

COMMON NAMES :

Moulmein rosewood, Jewels on a string (English); Mitoshika (Hindi); Tuma (Bengali)

This is a small to medium sized tree very similar to the *Milletia pinnata* (Pongamia), but not so common. It is a deciduous tree which blooms profusely with lavender, purple pea like flowers. When the tree is in bloom, it is leafless and one can only see the drooping clusters of flowers hang down giving the entire tree a purple look.

When not in bloom the tree is covered by the compound leaves. Small egg shaped, leaves about 8-10 inches are arranged along the midrib of the compound leaf. Bees throng the flowers for honey.

The timber is hard, heavy and is as useful as some other rose woods.

It is commonly grown for ornamental purposes in gardens as well as along roadsides. Even when planted in large containers, they bloom profusely.

Propagation is by seeds.

FAMILY
Fabaceae

NATIVE
Tropical and Temperate Asia

FLOWERING TIME
March to May

SCIENTIFIC NAME:

Milletia pinnata

COMMON NAMES :

Pongam tree, Indian beech tree (English); Karanj (Hindi); Hulagala, Honge (Kannada); Karanj (Marathi); Pungai (Tamil); Pungu (Telugu); Ponnu (Malayalam)

With the arrival of spring, the pongamia tree produces fresh green leaves and bunches of white-pink-mauve coloured flowers. The flowering is so profuse that the branches literally hang low due to the weight of the flowers. When in full bloom the white and pink flowers contrasting with the light green leaves is indeed a sight to treasure. New flowers each resembling the pea flower open up every day, simultaneously dropping off the old flowers carpeting the ground below. These flowers are collected by gardeners to make nutrient rich compost.

The flowers give way to oblong pods with a single seed. The seeds are traditionally prized for the oil content, which is used for lighting lamps, soap making and for lubrication. In recent times the seeds have gained importance as a potential source of bio-diesel and bio pestcide.

This tree has often been grown along field bunds and in homestead gardens for the useful seeds. However because of its medium size, good canopy and beautiful flowers, it has been used by city planners for planting along roads and in residential areas.

Propagation is by seeds.

FAMILY
Bignoniaceae

NATIVE
Burma

FLOWERING TIME
April-May and October-November

SCIENTIFIC NAME:

Millingtonia hortensis

COMMON NAMES :

Tree Jasmine, Indian cork tree (English); Neem Chameli (Hindi); Akash mallige (Kannada); Kat-malli (Tamil);Katesam (Malayalam)

On noticing a carpet of spangled flowers on the ground or the smell of pleasant fragrance, one can sense the Millingtonia tree. Abundant flowers are produced high up on this tall tree. Each flower has a pale green, slender tube ending in four petals with an equal number of stamens. The yellow coloured anthers are quite prominent. The waxy coating on the petals helps retain their freshness. The leaves are green for most part of the year and resemble those of the neem tree.

Children pick fallen flowers and use the tubular calyx to make whistles. The slender calyx also helps to braid the flowers into garlands without the use of thread. Certain parts of the tree are supposed to have medicinal value. The wood is light in weight and the bark provides an inferior quality cork.

The tree is straight and upright, growing as tall as 65-70 ft., mainly grown for its ornamental value. It does not help much as a shade tree. This tree is a good avenue tree, and for planting in parks and gardens. In some parts of the country flat and woody seed pods are seen hanging. However seed pods are not frequently seen.

Young plants grow from the dispersed seeds by wind and in the absence of seed-setting propagation is by root suckers

FAMILY
Sapotaceae.

NATIVE
South and South East Asia

FLOWERING TIME
June to December

SCIENTIFIC NAME:

Mimusops elengi

COMMON NAMES :

Spanish cherry, Bullet wood, Indian Medlar (English); Bakul, Maulsari (Hindi); Bakali, Ranjal (Kannada); Bakuli (Marathi); Bakul (Bengali); Ilanni (Malayalam)

The small sized flowers of about 5 cm in diameter are inconspicuous by themselves. These disc shaped creamy white flowers are borne in small clusters in the leaf axils. They open at twilight and gradually become very fragrant through the night, until they are shed next morning. The flower is prized for the fragrance it emits. In the morning thousands of these flowers are seen dotting the ground beneath the tree. When there is a light breeze or the tree is shaken the flowers sail to the ground looking like tiny UFOs (unidentified flying objects) floating down towards the earth. Leaves are small, dark green, smooth, shiny and oval shaped.

The tiny flowers are threaded to make garland for pujas or for adorning the hair. Considered as auspicious this flower is offered to Lord Ganesha during the 21 pushpa puja. They are also distilled to produce scented water.

The oval shaped fruits resembling the olive fruit are sweet when ripe and are edible. All parts of the plant are said to have medicinal properties and so the tree is in great demand in Ayurvedic medicine and modern pharmacological industry.

This medium sized evergreen tree of about 12-15 feet grows throughout India and is commonly planted along the roads. The dense, spreading crown makes it a good shade tree.

Propagation is via seed and stem cuttings.

FAMILY
Muntingiaceae

NATIVE
South Mexico to Bolivia

FLOWERING TIME
All round the year

SCIENTIFIC NAME:

Muntingia calabura

COMMON NAMES :

Strawberry tree, Jam tree (English); Gasagasi hannina mara (Kannada); Paanchara (Marathi); Nakkaraegu (Telugu)

The small tree with a short trunk produces white flowers that resemble those of strawberries. These white flowers about an inch in diameter dotting the green canopy are mostly hidden among the leaves. Each flower has 5 crinkled crepe petals with prominent yellow stamens. The flowers are short lived, blooming in the morning, only to fall off by afternoon.

The fruits which are produced all round the year are red-yellow about 1-1.25 cm in diameter. The soft pulp within the thin skin of the fruit is nauseatingly sweet which may or may not be relished by everyone. The fruit is filled with minute seeds that resemble the poppy seeds. The fruits are eaten fresh or prepared into a jam.

The tree is planted as a source of timber and fuel. Its soft wood is used for rural construction, while the bark is fibrous and used for making ropes.

This hardy and fast growing small tree has a beautiful umbrella like canopy. It is therefore popular for planting along roadside, in parks and gardens for ornamentation, shade and shelter. Though it provides good shade the tree is considered messy because of the falling of the sticky fruits especially when vehicles are parked below.

Propagation is from half-ripe wood cuttings and air layering rather than from seeds.

FAMILY
Apocynaceae

NATIVE
Southwest Asia

FLOWERING TIME
All round the year

SCIENTIFIC NAME:

Nerium oleander

COMMON NAMES :

Nerium or Oleander (English), Kanagile (Kannada), Kaner (Hindi), Raktakarabi (Bengali) and Arali (Tamil)

This year round blooming shrub or small tree lends colour to the garden and provides flowers for the daily puja. Flowers in shades of white, pink and red are borne in clusters at the tip of the branches. Each flower, either in single or double petals is about 1-2 inches in diameter. They possess a mild fragrance. The narrow and straight leathery leaves hold the bunch of flowers as in a bouquet.

It is a hardy plant which can be grown with little or no care. If grown as a shrub, it requires pruning at the end of the winter season to induce new growth & flowering and keep the plant in shape.

It is a very popular ornamental plant in private gardens, public parks and median strips of highways. While the taller varieties of plants can form hedges, the newer dwarf varieties can be grown as border plants. A word of caution is necessary, because all plant parts are poisonous and care should be taken to keep children away.

Plants can be multiplied by cuttings or can be purchased from nurseries.

FAMILY
Oleaceae

NATIVE
India and South-East Asia

FLOWERING TIME
June to December

SCIENTIFIC NAME:
Nycthanthes arbortristis

COMMON NAMES :
Coral Jasmine, Tree of sorrow (English); Har singar (Hindi); Parijata (Kannada); Shefali (Bengali); Pavizhamalli (Tamil); Paarijatam (Malayalam).

This flower, which does not see the light of the day blooms at dusk and carpets the ground below by dawn. The tree bereft of its flowers is thought to look sad and hence the common name-Tree of Sorrow. The flowers are white in colour with five to eight petals. The centre and the stalk are bright orange in colour. The pleasant fragrance of the flower pervades the surroundings. This flower has special significance in Hindu mythology and is considered auspicious for pujas.

The tree is small, grows to a height of 8-10 ft, and sheds its leaves in winter. The leaves are dark green when young and turn greyish with age. The fruit is flat, brown in colour with two sections each having a single seed.

This is probably the only flower which is picked up from the ground and yet considered fit for offering to the gods. The tree is common in many household gardens where the flowers are used for pujas or just kept in a bowl for the fragrance.

Propagation is through seeds or by cuttings.

In Indian mythology there are many stories surrounding this tree. It is said that it was Lord Krishna who brought this tree to the earth. However both his wives; Satyabhama and Rukmini wanted to have the tree, in their respective courtyards. Not wanting to displease either of them, Krishna planted the tree in Satyabhama's courtyard in a way that the flowers fell in Rukmini's courtyard.

Another story goes that a princess "Parijatak" who was in love with the sun tried to win his heart. On being rejected she committed suicide and a tree sprung from her ashes. Unable to stand the sight of the lover, the tree flowers only at night and sheds them like tear drops before the sun rises.

FAMILY
Salicaceae

NATIVE
South Africa

FLOWERING TIME
Late Spring and Summer

SCIENTIFIC NAME:

Oncoba spinosis

COMMON NAMES :

Snuff box tree. Fried egg tree (English)

The flowers of this tree resemble the *Mesua ferrea* or Nagkesar. The golden centre of the masses of yellow overlapping stamens surrounded by white petals make this a gorgeous flower. When the flowers fall flat facing upward, they resemble the fried egg, with the sunny side up. The singly borne flowers possess a honey like fragrance and attract a lot of butterflies. Individual flowers are short lived but bloom in succession over a period of time. The dark glossy green leaves are somewhat leathery with a small spine at the base

A slow growing, spiny tree about 12-15 ft. in height, may sometimes be considered a shrub. It is cultivated for its ornamental value and normally grown as a hedge plant. The tree is grown in Africa and India for its fruits

The hard shells of the dried fruits are used as snuff boxes. The dried fruits are used by children as tops for spinning and as rattles, because the loose seeds make a rattling noise, making them nature's toy for children to play with.

In its native of African countries various parts of the tree are used for medicinal purposes.

New trees can easily grown from the numerous seeds produced in the rounded fruits.

FAMILY
Malavaceae

NATIVE
Central and South America

FLOWERING TIME
Winter through Summer

SCIENTIFIC NAME:

Pachira aquatica

COMMON NAMES :

Malabar chestnut, French pea nut, money tree Provision tree (English)

Borne at the tip of the branches, the flowers are among the largest tree flowers. The long and narrow petals open up like a banana peels to reveal a host of hair like yellowish orange stamens. They bloom singly or in twos & threes. The flowers that open up at dusk last for less than a day. Being fragrant they attract a variety of birds, butterflies and bees. The upright unopened flower buds are shaped cigar like. The leaves are palmate, that is, with several finger like leaflets held together at the base.

Trees can grow as tall as 60 ft and is suitable for growing as an ornamental plant along roadsides, in gardens & parks. The tree is also grown in containers as a houseplant and often sold in nurseries with braided trunks. Although these container trees may not flower, the braiding of the trunks makes it an interesting house plant. The tree can also be trained as a bonsai.

In its native the tree has many medicinal uses.

The fruits are made up of five compartments, holding within them several seeds. The seeds or the nuts are light brown, striped white. They are tasty and consumed raw, roasted or ground into flour.

New trees are grown from seeds and cuttings.

FAMILY
Acanthaceaea

NATIVE
Eastern Africa

FLOWERING TIME
December- January

SCIENTIFIC NAME:

Parkia biglandulosa

COMMON NAMES :

Badminton ball tree (English); Chendul-ka-jhar (Hindi); Shivalinga -mara (Kannada); Chendu phul (Marathi)

This is a very interesting tree where each individual flower cannot be distinctly noticed as a flower. Hundreds of tiny flowers are borne on a ball like structure. During bloom time, globose fluffy balls hang down from branches of 1-2 ft. Initially, they are small bead like balls in rust colour. These gradually grow in size and the rust colour buds suddenly burst into white flowers. The entire ball like structure is covered with these creamish white flowers giving the look of a badminton ball; in shape, size and fluffy feel.

The tree is medium to small size, resembling that of the Gulmohar tree with its feathery compound leaves remaining green all through the year. After the flowers have faded and fallen off, small bean like flat seed pods are seen hanging in clusters at the tip of the stalk. The seeds are said to be edible in some communities.

This could be an interesting and attractive tree for avenue planting.

Propagation is by seeds.

FAMILY
Fabaceae

NATIVE
Southeastern Asia

FLOWERING TIME
March to May , September to December

SCIENTIFIC NAME:

Peltophorum pterocarpus

COMMON NAMES :

Copper pod, yellow flame tree (English); Haladi Gul Mohar (Hindi); Haladi Gulmohar (Kannada)

Come summer and this tree produces bright yellow flowers in abundance. The fragrant flowers, each of about 1-2 inches across are borne in clusters. The crinkled petals have a reddish brown mark on them. The flowers are followed by copper coloured seed pods, giving the plant its common name. During its peak bloom in summer, one can see a blanket of flowers on the wide canopy above and a carpet of flowers below. The flowers are followed by green seed pods which later turn to copper colour. The copper pods covering the canopy also gives a unique look to the tree

This tall tree growing up to 50 ft. in height, has become the most popular avenue tree for planting along roadsides. With a crown of up to 25 ft. across, it makes a very good ornamental and shade providing tree. It is often planted alternately with the red Gulmohar. Since both the trees bloom at the same time, the splash of colours is a wonderful sight.

Common method of propagation is by seed.

FAMILY
Apocynaceae

NATIVE
Central and South America

FLOWERING TIME
All round the year

SCIENTIFIC NAME:

Plumaria obtusa

COMMON NAMES :

Temple tree, Singapore grave yard flower, Deadman flower (English); Chameli, Gul-e-chin (Hindi); Kadusampige, Deva kangalu (Kannada); Khairchapha (Marathi)

This is a tropical growing, small tree, producing flushes of flowers almost all year round. There are many different species with variations in colour and size of the flowers and leaves. The most common of the Plumrias is the *P. obtusa* which produces clusters of fragrant white flowers with yellow centres. The bunches are in the form of a bouquet are borne on long stalks. Each flower has 5 petals and is about 3 inches wide. Flowers are followed by fruits which contain winged seeds.

This tree is as much prized for its leaves as it is for the flowers. The pure white flowers contrast beautifully with the dark green leaves. The leaves are leathery and shiny with prominent veins on the lower surface. While other species of *Plumerias* have pointed leaves, this one has rounded tips and hence the species name *obtusa* .This species also does not shed its leaves in winter and so remains green all through the year.

This is a FengShui plant suitable for planting in large gardens and as an avenue tree. Although the wonderful fragrance of the flowers attract bees and butterflies, they are disappointed because the flowers do not contain any nectar.

When propagation is by cuttings, flowering starts within a year. While seed grown trees take at least three years.

FAMILY
Sterulaceae

NATIVE
Southeast Asia

FLOWERING TIME
March- May

SCIENTIFIC NAME:
Pterospermum acerifolium

COMMON NAMES :

maple leafed Bayer, Dinner plate tree (English), Muchakuera (Hindi), Matesaleanda (Tamil), Lolugu (Telugu), Muchhakanda Champa (Bengali), Kanaka Champa (Kannada), Chitrak, Kshatra Vruksha

The white colored flowers of this tree are fragrant and nocturnal. They bloom at night with a pleasant fragrance. Flowers begin as one long bud, which then separate into five sepals, each of which is about 7 inches. Flowers are followed by brown elongated fruits which may take up to one year for maturity. The fruits hold many winged seeds.

The round shaped leaves are up to 12 inches in diameter with toothed edges. Being fairly wide they are used as leaf plates and hence the common name. In ayurveda it is an important medicinal plant for treating headaches, ulcers, wounds, cold, cough, bleeding disorders etc.

Grown mainly for ornamental purposes, it is planted along road sides and in large gardens and parks. The wide canopy provides good shade.

Propagation is from seeds.

FAMILY
Caesalpiniaceae

NATIVE
India, Burma, Malaya and Srilanka

FLOWERING TIME
Almost all year round

SCIENTIFIC NAME:
Saraca asoca

COMMON NAMES :

Asoka, Sorrowless tree (English); Sita Asoka (Hindi); Sita ashoka (Kannada)

The *Saraca asoca* produces attractive clusters of yellow-orange coloured flowers (resembling the *Ixora* flowers). These clusters containing numerous flowers are pinned closely to the main branches. The fresh blooms being yellow turn red with age. With heavy blooming from February – May, a few clusters are seen all-round the year. This is the state flower of Odisha.

New leaves are usually produced during spring time. The attractive bunches of reddish leaves are delicate and hang down limply.

All parts of the plant are useful in Ayurveda and more specifically the bark of the tree which is used in treating women's reproductive health problems. Over exploitation for medicinal value has rendered it an endangered species in most parts of the country.

This plant is considered important from the religious and cultural point of view in India, as it is said that Buddha was born under this tree. Although the plant had skipped the popular perception for planting in parks & gardens it is now gaining importance.

This is a small tree with horizontally spreading canopy and should not be confused with the popularly known false Ashoka *(Polyalthia longifolia)*, the tall, columnar tree grown as a wind breaker and a noise absorber.

Propagation is from seeds and stem cuttings.

FAMILY
Araliaceae

NATIVE
Australia

FLOWERING TIME
February-April

SCIENTIFIC NAME:

Schefflera actinophylla

COMMON NAMES :

Octopus tree, Queeensland umbrella tree (English)

At the top of this broad leaved tree, stand out the bright red spikes of flowers. These spikes or racemes resembling the tentacles of the octopus are about 2-3 ft in length containing upto a thousand flowers. The small, dull red flowers remain on the plant for about four months. The nectar filled flowers are foraged upon by bees and birds.The red blooms are followed by reddish- purple half inch fruits.

This evergreen tree is topped with a crown of large showy and attractive leaves. The name 'Umbrella Tree' is due to the rosette formation of the leaves. The species name *actinophylla* also means radiating leaves.

When grown in large open spaces the tree grows fairly tall, but can also be maintained in residential and institutional gardens.

Propagation is by seeds, cuttings or layers.

FAMILY
Fabaceae

NATIVE
Africa

FLOWERING TIME
Spring through Summer

SCIENTIFIC NAME:

Schotia brachepetalum

COMMON NAMES :

Weeping boer-bean, tree fuchsia, African greenheart and African walnut (English)

The red flowers begin to appear straight after the new leaves start emerging in Spring. The flowers produced in masses are rich deep red in colour. They are borne in dense branched heads on the old wood. The beauty of the flowers are the brightly coloured calyces (sepals), stamens and pedicels, while the actual petals are thin filaments which are quite insignificant The flowers are so rich in nectar that it literally drips down from the flowers. The "weeping" label in its common name refers to the copious quantities of nectar that may rain from the flowers when shaken. This attracts, bees, sun birds, monkeys and other insects

It is a medium to large tree (16- 50 ft.) with a wide-spreading, densely-branched, rounded crown. With its decorative foliage, showy flowers and symmetric bole, this is a good shade-providing tree for gardens and parks. However it is not recommended for planting over paved areas, car parks etc, because of the dripping nectar from the flowers.

The seeds rich in carbohydrates are consumed roasted or steamed and also substituted for coffee. Parts of the plant are used as medicines in local communities.

Propagation is by seeds and stem cuttings

FAMILY
Solanaceae

NATIVE
South America

FLOWERING TIME
All round the year in flushes

SCIENTIFIC NAME:

Solanum wrigtii

COMMON NAMES :

Potato flower (English); Mokoi (Hindi); Badane huvu (Kannada)

An ornamental plant belonging to the family of Brinjal, Tomato and Potato, produces five-lobed, star-shaped, purple-coloured flowers with prominent yellow anthers. When freshly opened, the flowers are dark purple, eventually fading to a pale lavender, nearing white. At any point of time one can notice 2-3 shades of the purple-lavender flowers on the plant. The fragrant flowers borne along new growth bloom in 3-4 flushes and continue on the plant over many weeks.. Flowering is better during the summer months than during other seasons.

Huge, large-lobed leaves, up to 1 ft long, with spines on the underside droop from its evergreen branches. Fruits are like small green balls, each of 3-4 cm in diameter containing numerous seeds within. Once the fruits mature and fall off, a new flush of flowers appear.

It is a fast growing woody perennial plant, which can be considered either a shrub or a small tree. Grown as a specimen tree, it is well suited for small gardens.

The flowers attract bees, butterflies, and birds. No parts of the plant should be ingested as they are poisonous.

Propagation is by cuttings and seeds. Flowering is faster when grown from cuttings.

FAMILY
Bignonaceae

NATIVE
Tropical dry forests of Africa

FLOWERING TIME
August- January

SCIENTIFIC NAME:

Spathodea campanulata

COMMON NAMES :

Fountain tree, African Tulip tree (English); Rugtoora (Hindi);Nirukavi (Kannada); Rudrapalash (Bengali); Patadi (Tamil)

During the flowering season the entire tree is covered with spectacular crimson red flowers as if set ablaze with fire. The tulip-shaped flowers with frilly edges are borne in bunches at the end of the branches. The cup-like flowers are ideal to hold the rain water. The nectar of the flowers mixed with the rain water attracts a variety of birds and other wild life. The tree has a long flowering season for nearly six months.

The interesting velvety, olive green, horn shaped buds are, filled with water. When the tip of the bud (horn) is pinched and simultaneously the broad end squeezed, water squirts out like a jet. This is the reason for the common names.

The tree attracts many species of birds. Since the wood is soft, it is used for nesting by hole building birds for nestling. Commercially the wood is important in the paper and plywood industry.

This is a fast growing tree starting to flower within 3-4 years of planting. Although it has gained popularity for planting in residential areas, the faded flowers which fall to the ground are considered to create a mess below the tree.

The tree is multiplied by seeds.

FAMILY
Bignoniaceae

NATIVE
South & North America

FLOWERING TIME
February March

Tabebuia roseo alba

SCIENTIFIC NAME:
Tabebuia Sps.

COMMON NAMES :
White trumpet flower (English)

Tabebuia is a genus of flowering plants native to the warm climates of North and South America. They are woody shrubs and trees, many with attractive flowers.

A few species produce timber, but the genus is mostly known for those that are cultivated as flowering trees and some species of *Tabebuia* are considered good for honey production.

In English *Tabebuias* have been called "trumpet trees", but this name is usually applied to other trees and has become a source of confusion and misidentification.

A few of the popular flowering ***Tabebuias*** grown in the city landscapes of the country are presented here

Tabebuia roseo alba: This one produces white coloured flowers in small bunches. Not only is the tree beautiful when in bloom, the falling flowers swathe the ground below in white.

FAMILY
Bignoniaceae

NATIVE
South America

FLOWERING TIME
February March

SCIENTIFIC NAME:

Tabebuia aurea

COMMON NAMES :

Caribbean Trumpet Tree, Yellow Tabebuia (English)

The tree shorn off its leaves in winter, comes to life in spring when the buds begin to appear. Bright yellow flowers in rounded bunches are formed at the tip of the branches. Each of the trumpet shaped flowers is 2-3 inches wide and equally long. A spectacular display of bright yellow flowers is seen on the leafless shoots. Though the entire tree blooms once a year, sporadic bunches of flowers may be seen all year round.

The trunk and branches of this small tree are twisted & contorted. The crown is also asymmetrical, giving the tree a picturesque and artistic look. Leaves are narrow, elongated and silvery colour.

This is an easy to grow tree which does not require much care and has now gained popularity. The major drawback of this tree however is that, the soft wood of the tree makes it susceptible to strong winds.

New trees can be grown from the numerous seeds produced in long pods

FAMILY
Bignonaceae

NATIVE
Argentina and Brazil

FLOWERING TIME
December-January

SCIENTIFIC NAME:

Tabebuia avellanedae

COMMON NAMES :

Pink Trumpet flower (English); Gulabi Tabubia (Marathi)

This is a spell binding tree when in full blossom. Being deciduous in nature, the tree is bereft of leaves when the buds begin to appear in clusters. The pink- magenta coloured flowers are borne in rounded clusters appearing like flower balls.

Each individual flower is trumpet shaped with a yellow throat about 2-3 inches long and 2 inches wide. The entire tree is in bloom for about 8-10 days after which the flowers begin to fall off, only to bloom the next year.

Soon after the flowers fall off the elliptical leaves with serrated margins begin to appear. Simultaneously the foot long seed pods containing several winged seeds also can be noticed on the tree.

In its place of origin the wood of this tree has many uses in addition to the ethno medicinal value of its various parts. However in India this tree has gained popularity for its ornamental value.

Propagation is by seeds.

FAMILY
Bignoniaceae

NATIVE
Tropical America

FLOWERING TIME
All year round

SCIENTIFIC NAME:

Tebubia berteroi

COMMON NAMES :

Hispaniolan rosy trumpet tree (English)

The tree bears flowers with 5 flaring petals with crinkled margins. The flowers are 5-6 cm long, rosy white with yellow in the throat. The flowering of this tree occurs when there are leaves in the tree and flowering is not as profuse as the other tebubias.

It makes a good avenue tree and is seen along roadsides.

Propogation is from seeds that are produced in long pods.

FAMILY
Bignoniaceae

NATIVE
Tropical America

FLOWERING TIME
January-February

SCIENTIFIC NAME:

Tabebuia umbellata

COMMON NAMES :

Golden trumpet tree (English); Basant rani (Hindi)

Not as common as the other Tabebuias this one also flowers when the leaves drop off for a short time. The tree is then covered in bright yellow, trumpet-shaped flowers. Although the colour of the flowers is exactly the same as that of *T aurea*, what identifies this species at first glance is the symmetrical and open form of the tree with horizontally spreading branches. Even when not in flower, it is handsome looking tree.

It is a small tree, deciduous for a short time in late winter. Shortly after flowering, it produces a new crop of five palmately divided leaflets.

Propagation is from semi ripe cuttings or air layering.

FAMILY
Bignoniaceae

NATIVE
Paraguay and Argentina (American tropics and subtropics)

FLOWERING TIME
January-February

SCIENTIFIC NAME:

Tabebuia rosea

COMMON NAMES :

Pink trumpet (English); rani (Hindi); Halka Gulabi Tabubia (Marathi)

There are many species of flowering Tebubias, but the prettiest among them is probably the *T. Rosea.* The entire tree looks like a huge bouquet when covered with baby pink flowers. When in bloom this tree has few equals. Each pink flower has a pale yellow centre. However the beauty of the tree lasts for a week only and one has to wait for a whole year to witness the spectacle.

During most of the year, the tree is covered with very characteristic broad leaflets that are pointed at the tip. Prior to bursting into bloom the tree sheds all its leaves and once the flowers wither, the leaves begin to grown again.

Although a small statured tree, the pink Tabebuia, can grow into a very big tree with the passing years. It is suitable for planting along roadsides, in large gardens and parks.

Propagation is from seed and flowering starts after 3-4 years.

FAMILY
Fabaceae

NATIVE
Tropical Africa

FLOWERING TIME
April-May

SCIENTIFIC NAME:

Tamarindus indica

COMMON NAMES :

Tamarind tree (English); Imli (Hindi) Hunasegida (Kannada); Chinch (Marathi); Amli (Bengali); Puli (Tamil)

The common name of Tamarind has its origin in the Arabic words "**Tamari Hindi**" meaning "**Fruit of India**". The tree is so ubiquitous in India that it is hard to believe that the tree is not native to India. The tree flowers profusely in summer presenting a kaleidoscopic sight, as though a host of butterflies has descended on the tree, are dangling from the branches. The flowers are borne in clusters, each being 2.5cm in diameter. Each flower has five petals, three of which are white and two colourful with pink/orange stripes. The three stamens resemble the antennae of the butterfly.

Flowers are sour in taste and are used for culinary purposes, while children also eat the raw tamarind pulp. In Ayurveda the flowers, leaves and fruits are known to possess medicinal properties. Commercially the tree is grown for the fruits as well as seeds. The sour fruit is an important ingredient in Indian cooking. The starch and gum from the seeds is used in many industries including textiles, printing, paper and pharmaceuticals. This is also a good tree for honey production.

It is a hardy tree commonly grown along farm bunds and roadsides from where the fruits are harvested. In recent times the tree is being cultivated as a commercial crop both: for the fruit and the seeds.

Propagation is by seeds.

FAMILY
Bignoniaceae

NATIVE
South America

FLOWERING TIME
June - July and October - November

SCIENTIFIC NAME:

Tecoma stans

COMMON NAMES :

Yellow bells, Yellow elder (English), Koranchelar (Kannada) and Piliya (Hindi)

This plant grown as a perennial shrub or small tree presents a colourful sight with bright yellow bell shaped flowers. Unlike most flowers which stand upright these fragrant flowers appear in drooping clusters at the end of the branches. Each flower is about 5cm wide with five wavy lobes. The throat inside is delicately etched with orange colour. The large leaves have 3-7 sharply pointed oval leaflets. The light but bright green colour of the leaves provides an attractive backdrop to the handsome yellow flowers.

It is commonly grown as a hedge plant in gardens 7 parks, or as a standalone tree. Pruning just after the flowering during monsoon and winter, promotes new flushes of flowers as well as controls shape and size of the plant. This hardy and easy to grow tree can be planted along the city roads.

The string bean like pods, contain flat oblong winged seeds. New plants sprout from this wind dispersed seeds. Plants can also be multiplied by semi-woody branch tips.

FAMILY
Lamiaceae

NATIVE
South and South East Asia

FLOWERING TIME
June to August

SCIENTIFIC NAME:

Tectona grandis

COMMON NAMES :

Teak (English); Sagun (Hindi); Saguvani (Kannada); Segun, Saigun (Bengali)

Although grown exclusively for quality timber, it is worthwhile to enjoy the beauty of the delicate flowers of the teak tree. The huge conical shaped bunches of flowers are seen at the tip of the branches. Most times these branches seem to be weighed down by the hundreds of tiny flowers. Each flower white in colour, is only about 6 mm, with a mild, and pleasant fragrance. While some flowers mature into seeds the others fall to the ground below forming a white lacy carpet.

Blooming in the rainy season, fruits ripen in winter and the tree stands bare from November to January. This deciduous tree occurs naturally in the warm tropical forests. Teak wood highly valued for its water resistance and durability, has gained popularity for growing on cultivated farms and residential gardens. The wood is used for boat building, house construction, furniture making, carving, etc.

Since seeds are hard to germinate the best way to plant these trees is by purchasing saplings from forest and private nurseries. The saplings may be seed sown or from tissue culture.

FAMILY
Malavaceae

NATIVE
India

FLOWERING TIME
July to February

SCIENTIFIC NAME:

Thespesia populnea

COMMON NAMES :

Umbrella tree, Portia tree, Indian Tulip (English);Paras papal (Hindi); Bugri gida (Kannada); Palaspipal (Bengali)

Resembling the tulip flower, the lemon- yellow coloured flowers look pretty among the dark green leaves. The five petals of the flower are very delicate, crinkled- crepe like. The fresh blooms are lemon yellow colour, which change to peach-pink the next day. They remain on the tree for three days before they fall off. Hence flowers of both the colours can be observed simultaneously on the plant. The flowers very much look like those of ladies finger and so the common name of *bhedi.*

The beautiful heart shaped leaves are green for most part of the year and are partly shed during early spring.

Flowers are followed by flat roundish pointed tipped (top like) seed pods. During summer when there are no flowers on the tree, one can see the brown seed pods hanging from the tree. The seed pods are shaped like tops, which children play with, by spinning them.

The wood of the tree is hardy and is used for building boats, making furniture and carts. Different parts of the tree have medicinal value. The oil extracted from seeds is used for lighting lamps and the leaves are used as green fodder.

This semi evergreen, small to medium sized tree is suitable for planting along roadsides to provide shade and beauty.

Propagation is by seeds.

FAMILY
Apocynaceae

NATIVE
South America

FLOWERING TIME
All round the year

SCIENTIFIC NAME:

Thevetia peruviana

COMMON NAMES :

Yellow Oleander (English); Peeli kaner (Hindi); Bitti (Marathi); Kolka-phul (Bengali)

The flowers of this small tropical tree come in three colours viz: yellow, white and peach. The yellow colour however is the most common. They are produced in clusters at the tip of the branches. The petals of this funnel shaped flowers are interesting, in that they are spirally twisted.

The evergreen tree with its spirally arranged, narrow and long waxy leaves is attractive even without the blooms.

Fruits are somewhat globular 4-5 cm. containing a single large seed. Initially green the fruits blacken on ripening.

It is an easy to grow plant evidenced by its presence in waste lands. It has also become popular as an ornamental plant in large gardens for border planting and for live hedges. It could make a good small tree for roadside and medians as it grows only 4-7 ft. in height and with a small crown.

It should be noted that all parts of the plant are poisonous especially the milky sap and the oily seeds. The common names, "be-still" and "suicide tree" refer to this characteristic of the plant.

The plant can be multiplied by seeds and hard wood cutting.

FAMILY
Apoynaceae

NATIVE
India and Burma

FLOWERING TIME
March -April

SCIENTIFIC NAME:

Wrightia tinctoria

COMMON NAMES :

Sweet Indrajao, Ivory wood, Pala indigo (English); Kapar, Dudhi (Hindi); Ajamara, Kodamurki, Aale/Beppale (Kannada); Kala kuda (Marathi); Paalai (Tamil)

It is a refreshing sight when one notices the delicate, white flowers, appearing like snowflakes among the freshly appeared bright green leaves. Blooming in summer, the flowers with five white petals, 2-3 cm long, turn creamish with age. When the flowers are in bloom, heavy bee activity is noticed all day long

This is a small, beautiful deciduous tree that grows up to 10 mtrs. Fruits are long brown follicles with flat hairy seeds within. As the fruit bursts open, the seeds are dispersed by wind.

All plant parts are known to possess medicinal properties. The leaves of this are a source of a blue dye called Pala Indigo. Being very light in weight the wood is used for making light weight articles. The famous wooden Channapatna toys of Karnataka are made from this wood. Leaves are considered a good source of green manure for paddy fields.

Propagation is from root suckers as well as seeds.